Lorenzo Giorgi

## A SIMPLE PHYSICS ÆNIGMA.

———

FIRST EDITION

———

*illustrated by Ornella Campanino*

1st edition.
Title: *A simple physicsænigma*.
© 2017 Copyright Lorenzo Giorgi.
© 2017 Illustrations by Ornella Campanino.
ISBN 978-1981995486.
**Any reproduction, representation, adaptation, translation and/or transformation, in whole or in part, is forbidden.**
If you wish to translate this book in a language other than Italian, e-mail me (please, attach Curriculum Vitæ): lg.universita@gmail.com

# SUMMARY

*Preface* ........................................................................... VII

*The problem* ...................................................................... 1

*Car J covers more distance than car W* ........................................ 3

*The two cars travel at the same mean speed
in every stretch of the road* ......................................................... 13

*Recapitulation* .................................................................. 15

*Solution* ........................................................................... 17

*Last remarks* ..................................................................... 19

## PREFACE.

How tedious is a long car journey spent alone!

How can one kill time? Some people listen to the radio. Some others prefer eating junk-food. I do not.

Usually, I simply wait for the ride to be over. And I get bored. A lot, poor me!

I drive twice a week between Grosseto and Pisa, two beautiful towns in the Italian Tuscany. I am a biologist, and a musician; surely not a physicist. However, when ennui attacks, my brain promptly fights back with a great variety of weapons, so that boredom is always taken by surprise and defeated.

That Friday was quite a windy day. The wind was blowing northward, while I was traveling southward, heading to Grosseto. Since I generally do not like to waste oil — to protect the environment, of course, not to save money... —, a windy day often means a low-speed Lorenzo, hence a long car journey; in other words, boredom. This time, my mind tried to break the siege through the power of Science.

I overtook a vehicle in a two-lane highway, then I went back to my own — which in Italy is the one on the right. I do not know exactly whether I inadvertently slowed down a little or the guy driving behind me speeded up, but I know that the guy's car matched my speed quite precisely, so that we traveled one after the other for some time, the distance between the two vehicles remaining fixed.

Then, the inspiring event: the road went downhill for some hundreds of meters, after which the slope disappeared. As soon as I could see the car behind me again, a physics problem popped up in my brain: "take that boredom", I thought, "you have been defeated once again!"

I am pleased to entertain the reader through the very same mental challenge that kept me company during that long car journey. May it cheer up your day! And here is a piece of advice: use it to challenge your friends, your students, your seminar audience, as well as anyone else who is brilliant enough to be amused by a mental challenge!

Lorenzo Giorgi.

# 1.

## THE PROBLEM

$W$ and $J$ are two cars traveling one after the other on a frictionless linear road. At the beginning of the observation, $W$ and $J$ are both moving in the same direction and at the same constant speed $k$. $W$ follows $J$ at a distance $d$:

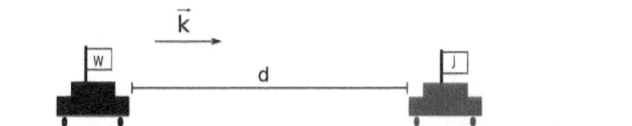

The road is horizontal for a distance $l$ ahead of $J$. Then the road goes downhill for a distance $h<d$; inclination is $\alpha$. Finally, the road is horizontal again:

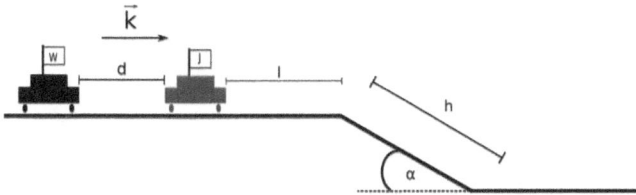

At the end of the observation, both cars have reached the second horizontal stretch of the road. *W* and *J* will be separated by a distance $\delta$:

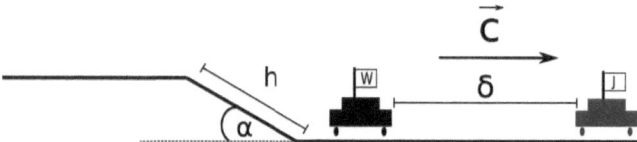

Acceleration is only due to gravity and confined in the downhill part of the road. We shall call *T* the total time of observation.

---

The paragraphs below will show that, at the end of the observation, *W* is still behind *J*. We will also show that the two cars have the same mean speed in the three parts of the road (higher horizontal stretch, downhill stretch, lower horizontal stretch). The time of observation is fixed. However, $\delta$ will be shown to be greater than *d*; in other words, *W* will have covered less distance than *J*. **Your job is to explain this paradox**.

## 2.

## CAR J COVERS MORE DISTANCE THAN CAR W.

Five phases will be distinguished:

- Phase 1 — from the beginning of the observation to the instant $J$ begins going downhill;

- Phase 2 — from the instant $J$ starts going downhill until it reaches the bottom of the inclined stretch of the road ($W$ is still traveling horizontally);

- Phase 3 — from the instant $J$ reaches the bottom to the instant $W$ begins going downhill;

- Phase 4 — $W$ goes downhill;

- Phase 5 — from the instant $W$ reaches the bottom to the end of the observation.

Phase 1 begins at instant $t_0$, Phase 2 begins at instant

$t_1$, etc. Observation ends at instant $t_5$. Furthermore:

$$\tau_1 = t_1 - t_0;$$
$$\tau_2 = t_2 - t_1;$$
$$\dots$$

So that:

$$\sum_{i=1}^{5} \tau_i = T$$

The variable $\theta$ will be employed. This variable is defined as:

$$\theta = t - t_b$$

where $t_b$ stands for the instant in which each phase begins. Thus, for example, during Phase 3:

$$\theta = t - t_2$$

In each Phase, $\theta$ varies between 0 and the $\tau$ value corresponding to that particular phase, i.e. the phase's duration.

The $\theta$ variable is necessary, because our analysis starts from calculating the distance $J$ and $W$ cover. Indeed, since:

$$v = \frac{ds}{d\theta}$$

the distance covered between two time points $t_n$ and $t_m$ will always be calculated as:

$$S = \int_{\theta_n}^{\theta_m} v d\theta$$

where:

$$\theta_n = t - t_n \wedge \theta_m = t - t_m$$

PHASE 1

Speed is constant for both cars:

$$v_J = v_W = k$$

Therefore, if $S_{J,1}$ is the distance covered by $J$ during Phase 1:

$$S_{J,1} = \int_0^{T_1} k d\theta = kT_1 - k \cdot 0 = kT_1$$

And of course:

$$S_{W,1} = kT_1$$

Hence:

$$\Delta S_1 = S_{J,1} - S_{W,1} = kT_1 - kT_1 \implies \boxed{\Delta S_1 = 0} \quad (1)$$

## PHASE 2

$J$ is accelerating. The acceleration $a$ is constant and related to gravity as follows:

$$a = g(\sin \alpha)^{-1}$$

The exact value of $a$ is not important. The important thing is that this value is known and easy to calculate. For the sake of simplicity, the acceleration will be henceforth termed $a$, instead of $g \sin \alpha$.

Now, the car's speed increases following the equation:

$$v = k + a\theta$$

Ergo:

$$S_{J,2} = \int_0^{\tau_2} (k + a\theta)d\theta = \frac{1}{2}a\tau_2^2 + k\tau_2 - a \cdot 0^2 - k \cdot 0 =$$
$$= \frac{1}{2}a\tau_2^2 + k\tau_2$$

$W$, instead, is still moving at a constant speed $k$. Therefore:

$$S_{W,2} = k\tau_2$$

Ergo:

$$\Delta S_2 = S_{J,2} - S_{W,2} = \frac{1}{2}a\tau_2^2 \implies \boxed{\Delta S_2 > 0} \quad (2)$$

## PHASE 3

Both cars are moving at a constant speed; indeed, $J$ has reached the maximum speed $c$, while $W$ is still traveling at the initial speed $k$. Ergo:

$$S_{J,3} = \int_0^{\tau_3} c\,d\theta = c\tau_3 - c \cdot 0 = c\tau_3$$

While:

$$S_{W,3} = \int_0^{\tau_3} k\,d\theta = k\tau_3 - k \cdot 0 = k\tau_3$$

While:

$$\Delta S_3 = S_{J,3} - S_{W,3} = c\tau_3 - k\tau_3 = (c-k)\tau_3$$

Ergo:

$$\boxed{\Delta S_3 > 0} \qquad (3)$$

## PHASE 4

$J$ is traveling at a constant speed $c$. Therefore:

$$S_{J,4} = \int_0^{\tau_4} c\,d\theta = c\tau_4 - c \cdot 0 = c\tau_4$$

W, instead, is accelerating. Its acceleration $a$ is constant and related to gravity as follows:

$$a = g(\sin \alpha)^{-1}$$

The car's speed increases following the equation:

$$v = k + a\theta$$

We will calculate the mean speed of car $W$ during Phase 4. Let us call this mean speed $x$. Then $x$:

$$S_{W,4} = x\tau_4$$

Ergo:

$$\Delta S_4 = S_{J,4} - S_{W,4} = c\tau_4 - x\tau_4 = (c - x)\tau_4$$

In particular:

$$c > x \implies \Delta S_4 > 0$$

Intuitively, $\Delta S_4 > 0$, because $c$ is the maximum speed reached by $W$ during Phase 4, hence $c > x$. We shall now demonstrate that $c > x$. The average value $\bar{y}$ of an integrable function $f(z)$ in an interval $[a; b]$ is:

$$\bar{y} = \frac{1}{b-a} \int_a^b f(z) dz$$

Hence:

$$x = \frac{1}{\tau_4 - 0} \int_0^{\tau_4} (k + a\theta) d\theta =$$
$$= \frac{1}{\tau_4}\left(k\tau_4 + \frac{1}{2}a\tau_4^2 - k \cdot 0 - \frac{1}{2}a \cdot 0^2\right) =$$
$$= \frac{1}{\tau_4}\left(k\tau_4 + \frac{1}{2}a\tau_4^2\right) = k + \frac{1}{2}a\tau_4$$

Our objective is to demonstrate that $c > x$. This is easy:

$$c > x \iff 2c > 2x \iff 2k + 2a\tau_4 > 2k + a\tau_4$$

Quod erat demonstrandum. We conclude that:

$$\boxed{\Delta S_4 > 0} \qquad (4)$$

## PHASE 5

Both cars travel at the same constant speed $c$:

$$v_J = v_W = c \implies S_{J,5} = S_{W,5} = c\tau_5$$

Ergo:

$$\Delta S_5 = S_{J,5} - S_{W,5} \implies \boxed{\Delta S_5 = 0} \qquad (5)$$

# CONCLUSION.

Let us consider the quantity:

$$\Delta S_{tot} = \Delta S_1 + \Delta S_2 + \Delta S_3 + \Delta S_4 + \Delta S_5$$

Clearly:

- if $\Delta S_{tot} > 0$, then $J$ covers more distance than $W$;

- if $\Delta S_{tot} = 0$, then $J$ covers the same distance as $W$;

- if $\Delta S_{tot} < 0$, then $J$ covers less distance than $W$.

Now, in light of equations (1) to (5):

$$\Delta S_{tot} = \underbrace{\Delta S_1}_{=0} + \underbrace{\Delta S_2}_{>0} + \underbrace{\Delta S_3}_{>0} + \underbrace{\Delta S_4}_{>0} + \underbrace{\Delta S_5}_{=0} \implies \Delta S_{tot} > 0$$

We conclude that car **$J$ covered more distance than car $W$** during the time of observation.

## 3.

## THE TWO CARS TRAVEL AT THE SAME AVERAGE SPEED IN EVERY STRETCH OF THE ROAD.

Let us subdivide the road into three stretches:

- Stretch U — the one before Stretch I;

- Stretch I — the inclined stretch of the road;

- Stretch D — the one after Stretch I.


### STRETCHES U AND D.

When a car is on Stretch U, it travels at speed $k$. The same goes for Stretch D. Therefore, **no matter what car is considered**, the mean speed on Stretch U is:

$$\bar{v}_U = k$$

Similarly, the mean speed on Stretch D is:

$$\bar{v}_D = c$$

for both cars.

## STRETCH I.

We have already calculated that, for both cars, the mean speed in Stretch I is:

$$x = k + \frac{1}{2} a \tau_4$$

Please note that:

$$\tau_4 = \tau_2$$

# 4.

# RECAPITULATION.

Time of observation is $T$. Car $J$ covers more distance than $W$. However, in each stretch of the road $J$ travels at the same mean speed as $W$. **How do you explain this paradox?**

# 5.

## SOLUTION.

In order to prevent an accidental look at the solution, I decided to condense it into a single sentence and encipher it. Each number corresponds to a letter:

1.2.   3.4.1.5.6.2.6.   7.

8.9.10.11.5.6.9.9.4.   10.

3.4.2.6.5.1.8.10.   12.   13.14.10.

15.9.6.15.6.9.16.1.6.14.4.   17.1.

8.4.18.15.6.   18.10.19.19.1.6.9.4.   9.1.11.15.4.8.8.6.

10.   20.

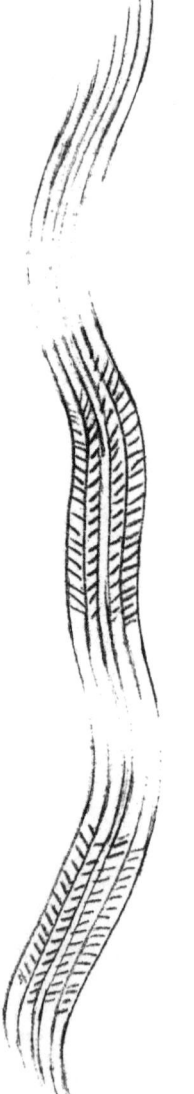

| | |
|---|---|
| A | 10. |
| C | 5. |
| D | 17. |
| E | 4. |
| G | 19. |
| I | 1. |
| J | 20. |
| K | 12. |
| L | 2. |
| M | 18. |
| N | 14. |
| O | 6. |
| P | 15. |
| R | 9. |
| S | 11. |
| T | 8. |
| U | 13. |
| V | 3. |
| W | 7. |
| Z | 16. |

# 6.

# LAST REMARKS.

This final section enumerates the variation allowed for each parameter. Indeed, although one might amuse himself by thinking about how the problem would degenerate when each parameter's value approaches (and even crosses) the limits that are enumerated here, I prefer to keep this problem real, so to speak, for an elegant solution to our paradox must be looked for inside the spectrum of a realistic variation. I am defying you with a real problem, not with a mere matter of extremes.

$$\begin{aligned} k &\in (0; +\infty) \\ d &\in (0; +\infty) \\ l &\in (0; +\infty) \\ h &\in (0; d) \\ \alpha &\in (0°; 90°) \end{aligned}$$

**ACKNOWLEDGMENTS.**

I thank my brother, Alberto Giorgi, for helping me with the book design, for encouraging me, and for his advice on the calculation of $x$.

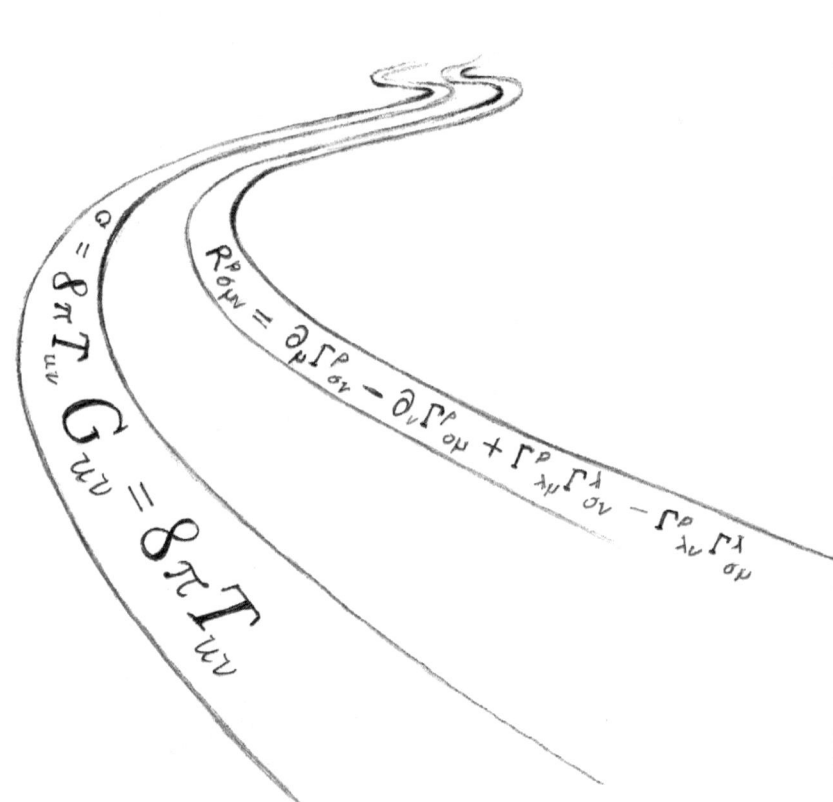

www.ingramcontent.com/pod-product-compliance
Lightning Source LLC
Chambersburg PA
CBHW031516210526
45464CB00007B/2936